# Problem Solving and Reasoning Pupil Book 2

Peter Clarke

William Collins' dream of knowledge for all began with the publication of his first book in 1819. A self-educated mill worker, he not only enriched millions of lives, but also founded a flourishing publishing house. Today, staying true to this spirit, Collins books are packed with inspiration, innovation and practical expertise. They place you at the centre of a world of possibility and give you exactly what you need to explore it.

Collins. Freedom to teach.

Published by Collins
An imprint of HarperCollins*Publishers*
The News Building
1 London Bridge Street
London
SE1 9GF

HarperCollins*Publishers*
Macken House, 39/40 Mayor Street Upper,
Dublin 1, D01 C9W8, Ireland

Browse the complete Collins catalogue at
**www.collins.co.uk**

© HarperCollinsPublishers Limited 2017

10 9 8 7 6 5 4

ISBN 978-0-00-826055-2

The author asserts his moral rights to be identified as the author of this work.

The author wishes to thank Brian Molyneaux for his valuable contribution to this publication.

British Library Cataloguing in Publication Data
A Catalogue record for this publication is available from the British Library

Author: Peter Clarke
Publishing manager: Fiona McGlade
Editor: Amy Wright
Copyeditor: Catherine Dakin
Proofreader: Tanya Solomons
Answer checker: Steven Matchett
Cover designer: Amparo Barrera
Internal designer: 2hoots Publishing Services
Typesetter: Ken Vail Graphic Design
Illustrator: Eva Sassin
Production controller: Sarah Burke
Printed and bound in the UK using 100% Renewable Electricity at CPI Group (UK) Ltd

# Contents

How to use this book      6

## Solving mathematical problems

Number –
Number and place value

| | |
|---|---|
| Multiples | 8 |
| 2-digit numbers | 9 |
| Ordering numbers | 10 |

Number –
Addition and subtraction

| | |
|---|---|
| Number grid | 11 |
| Stepping stones | 12 |
| Darts | 13 |
| Card calculations | 14 |

Number –
Multiplication and division

| | |
|---|---|
| Cuisenaire rods | 15 |
| Make a multiplication | 16 |
| Multiplying and dividing by 2 | 17 |
| Halving chains | 18 |

Number –
Fractions

| | |
|---|---|
| Four cubes | 19 |
| Fraction statements | 20 |
| Fraction wall | 21 |

Measurement –
Length
Mass
Capacity
Time
Money

| | |
|---|---|
| Giraffe order | 22 |
| 100 g to 1 kg | 23 |
| Making 1 litre | 24 |
| Ordering time | 25 |
| Buying sweets | 26 |

Geometry –
Properties of shapes

| | |
|---|---|
| Matchsticks | 27 |
| Sorting shapes | 28 |

Geometry –
Position and direction

| | |
|---|---|
| Making patterns | 29 |
| Turning | 30 |

Statistics

| | |
|---|---|
| Card sort | 31 |

## Reasoning mathematically

**Number –**
Number and place value

| | |
|---|---|
| Step counting | 32 |
| Which is larger? | 33 |
| Odd one out | 34 |

**Number –**
Addition and subtraction

| | |
|---|---|
| Hard or easy? | 35 |
| Same answer, different numbers | 36 |
| Always, sometimes, never | 37 |
| Spot the relation | 38 |

**Number –**
Multiplication and division

| | |
|---|---|
| 12 | 39 |
| Describing 24 | 40 |
| What's the number sentence? | 41 |
| Missing numbers | 42 |

**Number –**
Fractions

| | |
|---|---|
| Same and different | 43 |
| Who's right? | 44 |
| Who's wrong? | 45 |

**Measurement –**
Length
Mass
Capacity
Time
Money

| | |
|---|---|
| How long? | 46 |
| Fruit | 47 |
| Half a litre | 48 |
| What are the questions? | 49 |
| 67p | 50 |

**Geometry –**
Properties of shapes

| | |
|---|---|
| Shape symmetry | 51 |
| What's my shape? | 52 |

**Geometry –**
Position and direction

| | |
|---|---|
| Ronnie the Robot | 53 |
| Production line | 54 |

**Statistics**

| | |
|---|---|
| Different data | 55 |

## Contents

### Using and applying mathematics in real-world contexts

| | 2014 National Curriculum for Mathematics | | | | | | | | | |
|---|---|---|---|---|---|---|---|---|---|---|
| | Number – Number and place value | Number – Addition and subtraction | Number – Multiplication and division | Number – Fractions | Number – Money (£) | Measurement – Length (L), Mass (M), Capacity (C), Time (T), | Geometry – Properties of shapes | Geometry – Position and direction | Statistics | |
| Sorting dominoes | • | • | | | | | | | • | 56 |
| Visitors | • | | | | | | | | • | 57 |
| Phone numbers | • | • | | | | | | | | 58 |
| Clever birthday | | • | • | | | • (T) | | | | 59 |
| Phone pad patterns | • | • | • | | | | | | | 60 |
| Calculating dominoes | | • | • | | | | | | | 61 |
| Stamps | | • | | | | | | | | 62 |
| Body links | | | • | | | • (L) | | | | 63 |
| Walking | • | • | • | | | • (L) | | | | 64 |
| Holding hands | • | • | • | | | • (L) | | | | 65 |
| Balls | | | | | | • (L&M) | | | | 66 |
| Cooked and uncooked | | | • | | | • (M) | | | • | 67 |
| Tinned fruit | | | | | | • (M&C) | | | • | 68 |
| Orange drink | | | • | | | • (C) | | | • | 69 |
| TV times | | • | | | | • (T) | | | • | 70 |
| Paper clips | • | | • | | | • (£,L&T) | | | | 71 |
| Sharing £12 | | | | • | | • (£) | | | | 72 |
| £100 to spend! | | • | | | | • (£) | | | | 73 |
| Supermarkets | | | | | | | • | | • | 74 |
| Animal shapes | | | | | | | • | | | 75 |
| Envelopes | | | | | | | • | | | 76 |
| Apple patterns | • | | | • | | | | • | | 77 |
| Breakfast | • | | | | | | | | • | 78 |
| Class pet | • | | | | | | | | • | 79 |
| When you've finished, … | | | | | | | | | | 80 |

# How to use this book

## Aims

This book aims to provide teachers with a resource that enables pupils to:

- develop mathematical problem solving and thinking skills
- reason and communicate mathematically
- use and apply mathematics to solve problems.

## The three different types of mathematical problem solving challenge

This book consists of three different types of mathematical problem solving challenge:

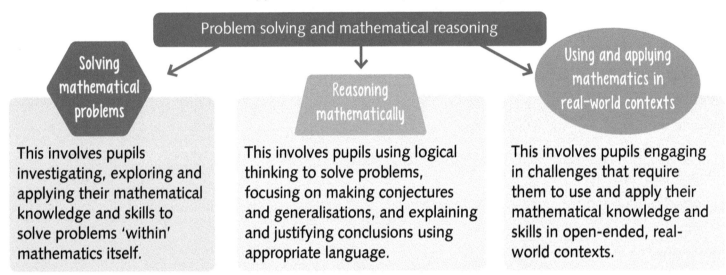

**Problem solving and mathematical reasoning**

**Solving mathematical problems**

This involves pupils investigating, exploring and applying their mathematical knowledge and skills to solve problems 'within' mathematics itself.

**Reasoning mathematically**

This involves pupils using logical thinking to solve problems, focusing on making conjectures and generalisations, and explaining and justifying conclusions using appropriate language.

**Using and applying mathematics in real-world contexts**

This involves pupils engaging in challenges that require them to use and apply their mathematical knowledge and skills in open-ended, real-world contexts.

This book is intended as a 'dip-in' resource, where teachers choose which of the three different types of challenge they wish pupils to undertake. A challenge may form the basis of part of or an entire mathematics lesson. The challenges can also be used in a similar way to the weekly bank of 'Learning activities' found in the *Busy Ant Maths* Teacher's Guide. It is recommended that pupils have equal experience of all three types of challenge during the course of a term.

The 'Solving mathematical problems' and 'Reasoning mathematically' challenges are organised under the different topics (domains) of the 2014 National Curriculum for Mathematics. This is to make it easier for teachers to choose a challenge that corresponds to the topic they are currently teaching, thereby providing an opportunity for pupils to practise their pure mathematical knowledge and skills in a problem solving context. These challenges are designed to be completed during the course of a lesson.

The 'Using and applying mathematics in real-world contexts' challenges have not been organised by topic. The very nature of this type of challenge means that pupils are drawing on their mathematical knowledge and skills from several topics in order to investigate challenges arising from the real world. In many cases these challenges will require pupils to work on them for an extended period, such as over the course of several lessons, a week or during a particular unit of work. An indication of which topics each of these challenges covers can be found on page 5.

## Briefing

As with other similar teaching and learning resources, pupils will engage more fully with each challenge if the teacher introduces and discusses the challenge with the pupils. This includes reading through the challenge with the pupils, checking prerequisites for learning, ensuring understanding and clarifying any misconceptions.

## Working collaboratively

The challenges can be undertaken by individuals, pairs or groups of pupils, however they will be enhanced greatly if pupils are able to work together in pairs or groups. By working collaboratively, pupils are more likely to develop their problem solving, communicating and reasoning skills.

## You will need

All of the challenges require pupils to use pencil and paper. Giving pupils a large sheet of paper, such as A3 or A2, allows them to feel free to work out the results and record their thinking in ways that are appropriate to them. It also enables pupils to work together better in pairs or as a group, and provides them with an excellent prompt to use when sharing and discussing their work with others.

An important problem solving skill is to be able to identify not only the mathematics, but also what resources to use. For this reason, many of the challenges do not name the specific resources that are needed.

## Characters

The characters on the right are the teacher and the four children who appear in some of the challenges in this book.

Mrs Edwards

Alice

Anushka

Matthew

Dylan

## Think about ...

All challenges include prompting questions that provide both a springboard and a means of assisting pupils in accessing and working through the challenge.

## What if?

The challenges also include an extension or variation that allows pupils to think more deeply about the challenge and to further develop their thinking skills.

## When you've finished, ...

At the bottom of each challenge, pupils are instructed to turn to page 80 and to find a partner or another pair or group. This page offers a structure and set of questions intended to provide pupils with an opportunity to share their results and discuss their methods, strategies and mathematical reasoning.

When you've finished, turn to page 80.

## Solutions

Where appropriate, the solutions to the challenges in this book can be found at *Busy Ant Maths* on Collins Connect and on our website: collins.co.uk/busyantmaths.

## Challenge

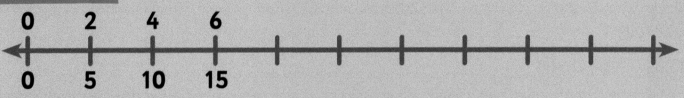

0   2   4   6

0   5   10   15

Draw this number line.

Starting from 0, write all the multiples of 2 above the line.

Starting from 0, write all the multiples of 5 below the line.

Now add together pairs of numbers that are above and below each other.

2 + 5 = ☐          4 + 10 = ☐          6 + 15 = ☐

What do you notice?

## Think about ...

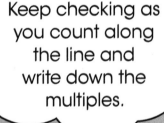

Keep checking as you count along the line and write down the multiples.

Think carefully how you're going to record the results of adding pairs of numbers that are above and below each other – you need to look out for any patterns.

## What if?

What if you find the difference between pairs of numbers that are above and below each other? What do you notice?

5 – 2 = ☐          10 – 4 = ☐          15 – 6 = ☐

What about adding together and finding the difference between pairs of multiples of 2 and 3?

What about adding together and finding the difference between pairs of multiples of 3 and 5?

When you've finished, turn to page 80.

## Challenge

0 1 2 3 4 5 6 7 8 9

**You will need:**
• set of 0–9 digit cards (optional)

Write one 2-digit number in each box. You can only use each digit once in a number.

| | Smallest 2-digit number | Largest 2-digit number |
|---|---|---|
| even number | | |
| odd number | | |
| multiple of 2 | | |
| multiple of 3 | | |
| multiple of 5 | | |
| multiple of 10 | | |

## Think about ...

What do you notice about the even numbers and the multiples of 2 you have written in the boxes? Why is this?

How do you know that a number is a multiple of 2, a multiple of 3, a multiple of 5 or a multiple of 10?

## What if?

Write 2-digit numbers in the boxes. You can only use each digit once in a number.

| | All the numbers between 40 and 50 | Number closest to 64 |
|---|---|---|
| even number | | |
| odd number | | |
| multiple of 3 | | |
| multiple of 5 | | |
| multiple of 10 | | |

When you've finished, turn to page 80.

## Challenge

34 < ☐ < 38

What numbers could you write on the red card?

Write a rule for finding all the numbers that you could write on the red card.

## Think about ...

There is more than one number: can you find all the possible numbers?

Check your numbers to make sure that they fit the statement.

## What if?

What if the cards are arranged like this?

49 > ☐ > 42

Or like this?

81 > ☐ < 50

What about like this?

67 < ☐ > 73

When you've finished, turn to page 80.

Addition and subtraction

## Challenge

Choose any square of four numbers from the grid.

Add together the two numbers in the opposite corners.

| 1 | 2 |
|---|---|
| 7 | 8 |

1 + 8 =

2 + 7 =

What do you notice?

| 1 | 2 | 3 | 4 | 5 | 6 |
|---|---|---|---|---|---|
| 7 | 8 | 9 | 10 | 11 | 12 |
| 13 | 14 | 15 | 16 | 17 | 18 |
| 19 | 20 | 21 | 22 | 23 | 24 |
| 25 | 26 | 27 | 28 | 29 | 30 |
| 31 | 32 | 33 | 34 | 35 | 36 |

## Think about ...

Try different squares of four numbers from the grid.

Make sure that you try enough 2 by 2 squares to be able to describe what you notice.

## What if?

What do you notice when you find the difference between the two numbers in the opposite corners?

What do you notice when you add or subtract the two numbers in the opposite corners on these number grids?

| 1 | 2 | 3 | 4 | 5 |
|---|---|---|---|---|
| 6 | 7 | 8 | 9 | 10 |
| 11 | 12 | 13 | 14 | 15 |
| 16 | 17 | 18 | 19 | 20 |

| 1 | 2 | 3 | 4 |
|---|---|---|---|
| 5 | 6 | 7 | 8 |
| 9 | 10 | 11 | 12 |
| 13 | 14 | 15 | 16 |

When you've finished, turn to page 80.

What about a 1–100 number grid?

## Challenge

Investigate the different totals you can make by stepping on the stones and adding up the numbers as you walk from **Start** to **End**.

## Think about ...

You can't step over any stones and miss them out.

You need to show what stones you step on to make a total.

## What if?

Moving from **Start** to **End**:

- What is the greatest total you can make without stepping on a stone more than once?
- What is the greatest total you can make by stepping on five numbered stones?
- What is the smallest total you can make?

Alice says: I got a total of 40.

Which stones did Alice step on?

When you've finished, turn to page 80.

## Challenge

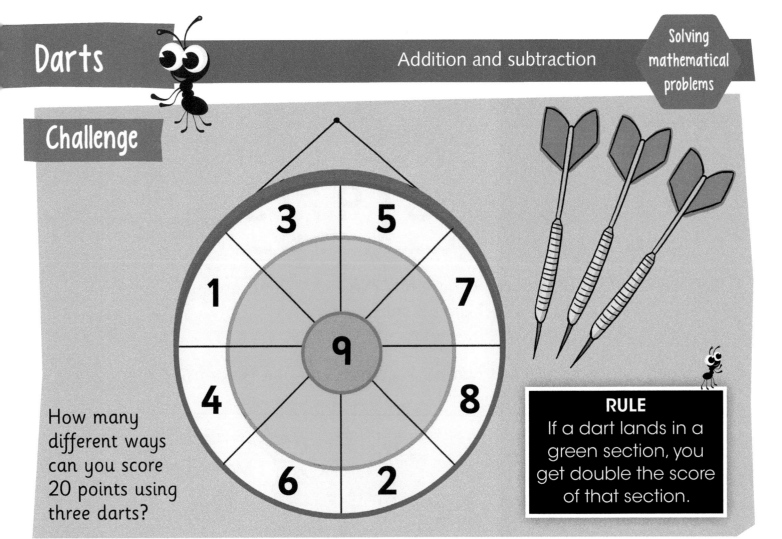

How many different ways can you score 20 points using three darts?

**RULE**
If a dart lands in a green section, you get double the score of that section.

## Think about ...

How are you going to record the different ways you can score 20?

Find ways of scoring 20 where one or more of the darts lands in the green section.

## What if?

What's the smallest points total you can score using three darts?

What's the largest points total you can score using three darts?

What other points totals can you score using three darts?

When you've finished, turn to page 80.

## Challenge

**8 9 3**

**You will need:**
• set of 1–9 digit cards

Using a set of 1–9 digit cards, choose any three cards and arrange them to make a 2-digit number and a 1-digit number.

Add the two numbers together.

How many 2-digit add 1-digit number sentences can you make using the three cards?

## Think about ...

Be organised so that you can find all the different number sentences possible.

Make sure that you check the answers to your number sentences.

## What if?

Choose three different digit cards and investigate what other 2-digit add 1-digit number sentences you can make.

What if you subtract the 1-digit number from the 2-digit number?

What if you choose four cards and arrange them to make two 2-digit numbers? Investigate what different addition and subtraction number sentences you can make.

**5 6   7 1**

When you've finished, turn to page 80.

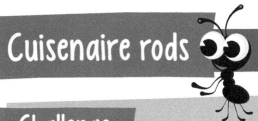
## Challenge

Use the Cuisenaire rods on the right to write a multiplication or division number sentence to describe each row of Cuisenaire rods below.

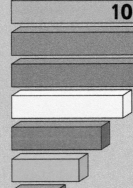

**10**

**2**

**You will need:**
• Cuisenaire rods

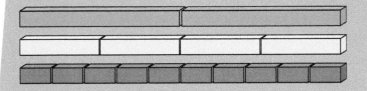

What about this set of Cuisenaire rods?

## Think about ...

What is the value of each of the Cuisenaire rods?

Can you write a multiplication and division number sentence to describe each row of Cuisenaire rods?

## What if?

Write a multiplication or division number sentence to describe each row of Cuisenaire rods on the right.

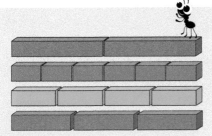

What about this set of Cuisenaire rods?

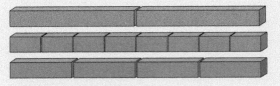

Using a set of Cuisenaire rods, create a wall similar to those above. Show your wall to a friend and ask them to describe the wall, using multiplication and division number sentences.

When you've finished, turn to page 80.

## Challenge

×   =

Use these numbers and signs to make multiplication number sentences.

2 × 4 =

How many different multiplication number sentences can you make?

## Think about ...

Make sure that you also work out the answer to each number sentence.

Try to find a system so that you can write as many number sentences as possible.

## What if?

What division number sentences can you make if a division card replaces the multiplication card?

How many different division number sentences can you make?

10 ÷ 2 =

When you've finished, turn to page 80.

## Challenge

Using a set of 0–9 digit cards, investigate making pairs of numbers where one number is double the other number.

**You will need:**
- set of 0–9 digit cards

## Think about ...

You can't use the same digit card twice when making pairs of numbers, such as

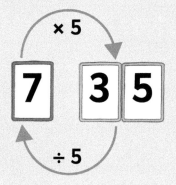

Think about recording your results as multiplication and division number sentences.

## What if?

What if you can use the same digit card twice when making pairs of numbers?

What about using the cards to make pairs of numbers so that one number is five times the other number?

When you've finished, turn to page 80.

# Halving chains

## Challenge

This is a 3-link chain:

Choose a 2-digit even number and keep halving until you reach an odd number.

What is the longest chain you can make, starting with a 2-digit number?

## Think about ...

How are you going to record your chains?

What patterns can you see in the numbers that make long chains? Can you use this to make other chains?

## What if?

What if you start with a 3-digit number?

What might a doubling chain look like?
What numbers could you start with?

When you've finished, turn to page 80.

# Four cubes

## Challenge

Using red and blue interlocking cubes, join four cubes together to make a shape.

**You will need:**
- red, blue and green interlocking cubes

How many cubes are red? How many are blue?

What fraction of your shape is red? What fraction is blue?

How many different ways can you make a shape using other combinations of four red and blue cubes?

How many cubes are red? Blue?

What fraction of your shape is red? Blue?

## Think about ...

Think about using the **number** of cubes of each colour to help you describe the shapes as **fractions**.

Remember that a fraction can be used to describe part of a group of objects.

## What if?

What if you join four red, blue and green interlocking cubes together to make a shape?
How many cubes are red? Blue? Green?
What fraction of your shape is red? Blue? Green?

What if you make a shape using six red and blue cubes?
How many cubes are red? Blue?
What fraction of your shape is red? Blue?

When you've finished, turn to page 80.

19

## Challenge

2 out of 5, or two-fifths, are wearing glasses.

How many different fraction statements can you make to describe the people in the picture?

## Think about ...

Use fractions and words such as 'out of' to describe what you see in the picture.

Think about what is the same and what is different about the people in the picture, and how many of each there are.

## What if?

What statements can you make about this tray of gingerbread?

When you've finished, turn to page 80.

# Fraction wall

## Challenge

| 1 whole | | | | | | | | | | | |
|---|---|---|---|---|---|---|---|---|---|---|---|

The fraction wall shows:
- 1 whole
- $\frac{1}{2}$, $\frac{1}{2}$
- $\frac{1}{3}$, $\frac{1}{3}$, $\frac{1}{3}$
- $\frac{1}{4}$, $\frac{1}{4}$, $\frac{1}{4}$, $\frac{1}{4}$
- $\frac{1}{5}$, $\frac{1}{5}$, $\frac{1}{5}$, $\frac{1}{5}$, $\frac{1}{5}$
- $\frac{1}{6}$, $\frac{1}{6}$, $\frac{1}{6}$, $\frac{1}{6}$, $\frac{1}{6}$, $\frac{1}{6}$
- $\frac{1}{8}$, $\frac{1}{8}$, $\frac{1}{8}$, $\frac{1}{8}$, $\frac{1}{8}$, $\frac{1}{8}$, $\frac{1}{8}$, $\frac{1}{8}$
- $\frac{1}{10}$, $\frac{1}{10}$, $\frac{1}{10}$, $\frac{1}{10}$, $\frac{1}{10}$, $\frac{1}{10}$, $\frac{1}{10}$, $\frac{1}{10}$, $\frac{1}{10}$, $\frac{1}{10}$
- $\frac{1}{12}$, $\frac{1}{12}$, $\frac{1}{12}$, $\frac{1}{12}$, $\frac{1}{12}$, $\frac{1}{12}$, $\frac{1}{12}$, $\frac{1}{12}$, $\frac{1}{12}$, $\frac{1}{12}$, $\frac{1}{12}$, $\frac{1}{12}$

Look at the fraction wall. How many different ways can you find of writing $\frac{1}{2}$?

Can you find a different way of writing $\frac{1}{4}$?

What rules can you make for finding out what fractions are the same as $\frac{1}{2}$ or $\frac{1}{4}$?

## Think about ...

How are you going to write your answers?

Can you think of other fractions that are the same as $\frac{1}{2}$ or $\frac{1}{4}$, without using the fraction wall?

## What if?

Anushka says:

I can see by the wall that 2 halves are the same as 1 whole.

What other fractions can you see on the wall that are the same as 1 whole?

What rule can you make for finding out what fractions are the same as 1 whole?

When you've finished, turn to page 80.

## Challenge

Gilmore is taller than Molly.
Tula is shorter than Bala.
Bala is almost as tall as Molly.

Write the names of the giraffes in order of height, starting with the shortest.

Now use the < and > signs to make statements comparing the heights of the giraffes.

## Think about ...

Read the clues several times and in different orders.

Think about using the < and > signs to compare two and more than two giraffes.

## What if?

Alice says:

The tallest giraffe is 6 m tall.
One giraffe is 1 m taller than the shortest giraffe.
The shortest giraffe is 4 m shorter than the tallest giraffe.
One giraffe is 1 m shorter than the tallest giraffe.

What is the height of each giraffe?

When you've finished, turn to page 80.

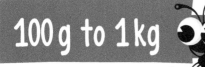

## Challenge

**500 g**
**500 g**
**500 g**
**500 g**

200 g
200 g
200 g
200 g

100 g
100 g
100 g
100 g

Using only 100 g, 200 g and 500 g weights, I can make all of these weights.

100 g
200 g
300 g
400 g
500 g
600 g
700 g
800 g
900 g
1 kg

100g
200g
300g
400g
500g
600g
700g
800g
900g
1 kg

Is Matthew right?

## Think about ...

Try making each weight, using the smallest number of 100 g, 200 g and 500 g weights.

Why are 100 g, 200 g and 500 g useful standard weights?

## What if?

Matthew also says:

In fact, using only 100 g, 200 g and 500 g weights, I can make some of these weights in more than one way.

Can you?

When you've finished, turn to page 80.

# Making 1 litre

## Challenge

Mrs Edwards says:

> I can find pairs of cylinders to mix together so that they each make 1 litre.

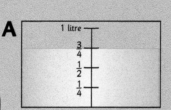
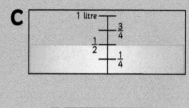
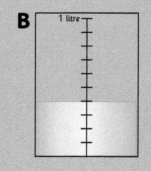
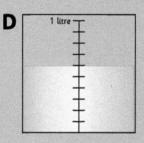
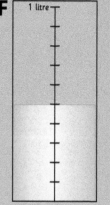
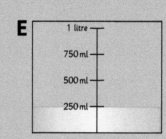

Which different pairs of cylinders should Mrs Edwards mix together?

## Think about ...

Think about how you're going to make 1 litre when some of the cylinders are labelled with fractions and others are labelled with numbers.

How are you going to show in your recording how you mixed different cylinders?

## What if?

Mrs Edwards also says:

What other amounts of liquid can you make by mixing two cylinders together?

> If I mix cylinders C and E together, I will get 750 ml of liquid.

When you've finished, turn to page 80.

# Ordering time

## Challenge

Look at these clocks.

What might you be doing at each of these times?

Write the times in order.

## Think about ...

Remember the different parts of the day – morning, afternoon, evening and night.

Why, when you compare what you have done with someone else, might you have ordered the times differently?

## What if?

Can you think of other things you might do at these times? Think of things where the time of the day is different from what you wrote before.

Write these times in order.

Why does this order still make sense?

When you've finished, turn to page 80.

## Challenge

Dylan loves sweets.

Each week he gets £1 to spend on sweets.

He always buys some of each of these six sweets.

How many of each of the six different sweets can Dylan buy with his £1?

Chews 3p

Gob stoppers 20p

Bananas 5p

Milk bottles 2p

Raspberries 1p

Liquorice 10p

## Think about ...

Remember that Dylan buys some of each sweet.

Can you think of more than one way that Dylan can spend his £1?

## What if?

One week, Dylan decides he wants to spend his £1 on only four different sweets.

How many of each of the four different sweets can Dylan buy with his £1?

When you've finished, turn to page 80.

## Challenge

17 matchsticks have been used to make 6 small squares:

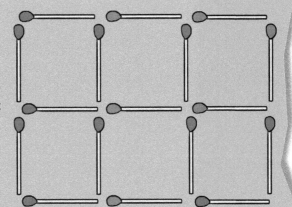

Remove 2 matchsticks to leave 5 small squares.

Remove 3 matchsticks to leave 4 small squares.

Remove 4 matchsticks to leave 3 squares. All the squares are **not** the same size.

Remove 9 matchsticks to leave 2 small squares.

**You will need:**
- pile of matchsticks
- squared paper or square dot paper

## Think about ...

All of your shapes must be squares and you can't have parts of a square:

Think carefully about how you're going to record your results to show how many matchsticks you have removed.

## What if?

18 matchsticks have been used to make 9 small triangles:

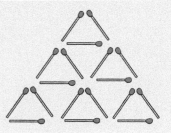

Remove 3 matchsticks to leave 6 triangles. All the triangles are the same size.

Remove 6 matchsticks to leave 4 triangles. All the triangles are **not** the same size.

Remove 9 matchsticks to leave 2 triangles. The triangles are **not** the same size.

When you've finished, turn to page 80.

## Challenge

Draw tables or diagrams to sort these shapes in as many different ways as you can. Each time, explain how you have sorted the shapes.

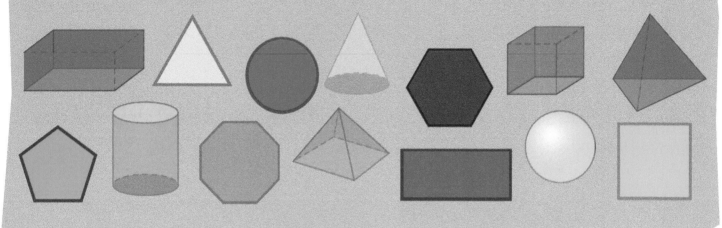

## Think about ...

What is the same about the shapes in each group?

Can you sort the shapes into two groups? What about three or four groups?

## What if?

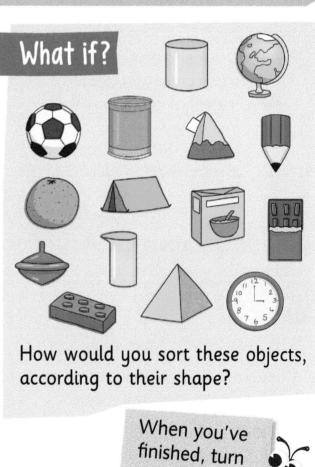

How would you sort these objects, according to their shape?

When you've finished, turn to page 80.

## Challenge

Place 12 counters of two or more colours on the circles below to make a repeating pattern.

Record your pattern.

How many different patterns can you make?

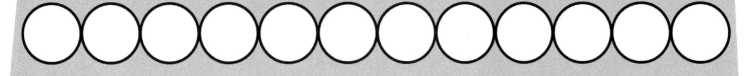

**You will need:**
- small counters in different colours
- coloured pencils

## Think about ...

Make sure that your pattern repeats at least once.

Don't use any more than four different colours in your pattern.

## What if?

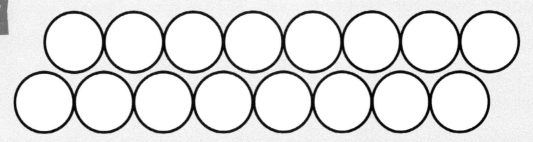

Place 16 counters of two colours on the circles above to make a repeating pattern.

What about using 16 counters of three different colours?

Show all of your patterns to a friend and ask them to continue your patterns.

When you've finished, turn to page 80.

## Challenge

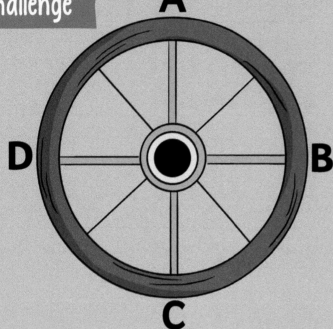

You can move from A to B by making a $\frac{1}{4}$ turn to the right.

You can move from B to D by making a $\frac{1}{2}$ turn clockwise.

Investigate different turns that go between two points on the wheel.

## Think about ...

Think about how you are going to record the different turns. Use words such as: **whole**, **half**, **quarter**, **three-quarter**, **right**, **left**, **right angle**, **clockwise** and **anticlockwise**.

Work in a system to try and find all the different turns possible.

## What if?

Alice also says:

I know another way you can move from A to B.

Can you find different ways to describe moving between two points on the wheel? Which description requires the smallest amount of turn?

When you've finished, turn to page 80.

## Challenge

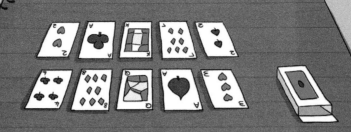

**You will need:**
- pack of playing cards

How many different ways can you sort a pack of playing cards?

## Think about ...

Be sure to sort all of the cards.

Think carefully about how you're going to record the different ways you sort the cards.

## What if?

Sort the cards using a Venn diagram. Can you create more than one Venn diagram?

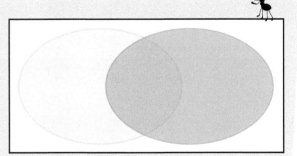

What about sorting the cards using Carroll diagrams?

When you've finished, turn to page 80.

## Challenge

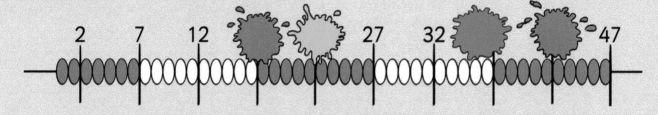

2  7  12  27  32  47

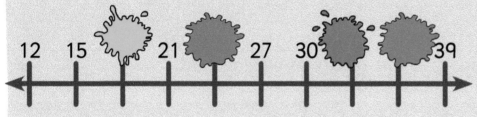

12  15  21  27  30  39

What numbers are covered by the ink splats?

Explain how you know.

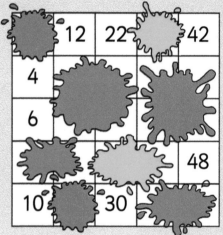

| | 12 | 22 | | 42 |
| 4 | | | | |
| 6 | | | | |
| | | | | 48 |
| 10 | 30 | | | |

## Think about ...

Think about counting in steps of the same size.

What patterns do you notice when you count?

## What if?

Alice says:

Is Alice right?

Explain why.

When I count in steps of 10 from a number, the ones digit stays the same.

When you've finished, turn to page 80.

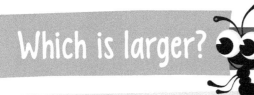
## Challenge

6 tens and 8 ones is larger than 5 tens and 19 ones.

No! 6 tens and 8 ones is smaller than 5 tens and 19 ones.

Who's right – Anushka or Dylan?

Explain your answer.

## Think about ...

Include representations when explaining your answer.

Is there another way to express 19 ones?

## What if?

Dylan also says:

Is Dylan right?

Explain why.

4 tens and 28 ones is the same as 6 tens and 8 ones.

When you've finished, turn to page 80.

## Challenge

Which of these arrows does **not** point to the number 63?

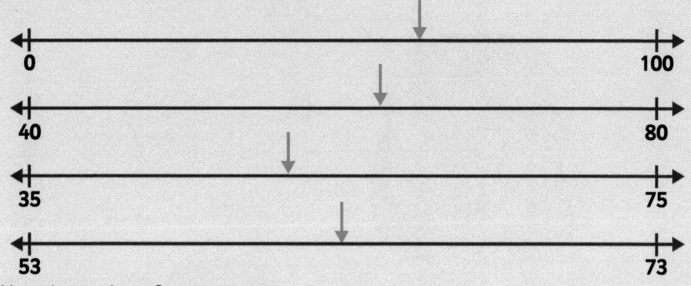

0               100

40               80

35               75

53               73

How do you know?

Draw the number line that is the odd one out and draw an arrow to show where the number 63 should be.

## Think about ...

Think about what number is half way along the number line.

Think about marking and labelling regular divisions along your number line.

## What if?

Draw a different number line. Mark the start and end numbers and draw an arrow pointing to 40.

Draw another number line with different start and end numbers and show where 40 should be.

When you've finished, turn to page 80.

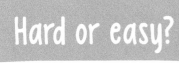
## Challenge

Arrange these number sentences into two groups: **hard** and **easy**.

| 32 + 7 = ☐ | 25 + ☐ = 33 | 78 – 34 = ☐ |

| 61 – 20 = ☐ | 45 – 8 = ☐ | 55 + 30 = ☐ |

| 8 + 7 + 3 = ☐ | 26 + 49 = ☐ | ☐ – 16 = 40 |

Explain why you find the **hard** number sentences more difficult.

Why do you find the **easy** number sentences easier?

## Think about ...

Which number sentences can you do quickly in your head, and which ones do you need to write down and think about more?

Can you spot any patterns in your group of **hard** number sentences? What about your group of **easy** number sentences?

## What if?

Write two different **easy** number sentences.

Write two different **hard** number sentences.

What's the hardest number sentence you can think of?

When you've finished, turn to page 80.

## Challenge

Write a number in each box to make a true statement.

4☐ + 2☐ = 71

How many different ways can you complete the number sentence?

Can you find them all?

What patterns do you notice?

## Think about ...

Hint: There are between five and ten different ways to make 71.

Look for patterns – this will help you to see if you have found all of the possible number sentences.

## What if?

What about these number sentences? Can you find all of the different number sentences possible?

What patterns do you notice?

8☐ − 3☐ = 57

☐4 + ☐5 = 99

☐3 − ☐7 = 26

When you've finished, turn to page 80.

## Challenge

When you add two odd numbers together, the answer is an even number.

When you add any three numbers, the answer is an odd number.

When you add an odd number and an even number, the answer is even.

Is each of Matthew, Dylan and Alice's statements always true, sometimes true or never true?

Provide examples to back up your answer.

## Think about ...

Make a list of the first five odd numbers and the first five even numbers to help you.

Make sure you provide enough examples to make your decision convincing.

## What if?

Matthew also says:

If you subtract an even number from another even number, the answer is an odd number.

Dylan also says:

If you find the difference between an even number and an odd number, the answer is an odd number.

Alice also says:

If you subtract an odd number from another odd number, the answer is an even number.

Is each statement always true, sometimes true or never true?

Provide examples to back up your answer.

When you've finished, turn to page 80.

## Challenge

| 7 | 9 | 16 | 17 | 19 | 26 | 36 | 70 | 90 | 160 |

Three of the number cards above have been used to make this addition number sentence.

$$7 + 9 = 16$$

Use the number cards to make as many different addition number sentences as you can.

Now look at the set of number sentences you have written. What patterns do you notice? What is the same about them?

## Think about ...

Think about how one number sentence can help you write other number sentences.

What stays the same and what changes between the numbers in different number sentences?

## What if?

Use the cards to make as many different subtraction number sentences as you can.

Then look at the set of addition number sentences and the set of subtraction number sentences you have written. What patterns do you notice? What's the same and what's different about the sets of addition and subtraction number sentences?

When you've finished, turn to page 80.

## Challenge

Choose number cards from above to complete these two number sentences.

$$\boxed{\phantom{0}} \times \boxed{\phantom{0}} = \boxed{12} \qquad \boxed{12} \div \boxed{\phantom{0}} = \boxed{\phantom{0}}$$

How many different multiplication and division number sentences can you make?

What do you notice?

## Think about ...

What patterns do you notice in the multiplication number sentences? What about the division number sentences?

Is there a relationship between the multiplication and the division number sentences?

## What if?

What if you use the number cards to complete these two number sentences?

$$\boxed{\phantom{0}} \times \boxed{\phantom{0}} = \boxed{20} \qquad \boxed{20} \div \boxed{\phantom{0}} = \boxed{\phantom{0}}$$

What about these two number sentences?

$$\boxed{\phantom{0}} \times \boxed{\phantom{0}} = \boxed{24} \qquad \boxed{24} \div \boxed{\phantom{0}} = \boxed{\phantom{0}}$$

When you've finished, turn to page 80.

## Challenge

Which of these number sentences does **not** describe the array?

Prove it.

| | |
|---|---|
| $2 \times 6 + 2 \times 6 = 24$ | $4 \times 6 = 24$ |
| $6 + 6 + 6 + 6 = 24$ | $6 \times 4 = 24$ |
| $4 \times 5 + 5 = 24$ | $24 \div 4 = 6$ |
| $4 + 4 + 4 + 4 + 4 + 4 = 24$ | $24 \div 6 = 4$ |
| $2 \times 6 + 6 + 6 = 24$ | $5 \times 4 + 4 = 24$ |

## Think about ...

You could use a diagram to prove which number sentence does not describe the array.

Imagine different rectangles within the array. How does this help you see different number sentences?

## What if?

How many different number sentences can you write to describe this array?

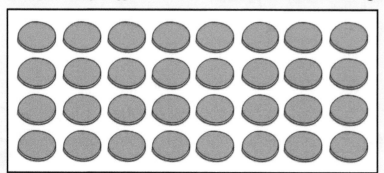

When you've finished, turn to page 80.

## Challenge

**Alice**

**Anushka**

**Matthew**

**Dylan**

| 5 | 5 | 5 | 5 | 5 | 5 | 5 |
|---|---|---|---|---|---|---|

Alice, Anushka, Matthew and Dylan were each asked to show a representation of a different number sentence.

What number sentence might they each have been asked to show?

Can you think of a different number sentence that they each might have been asked to show?

## Think about ...

Does the representation show a multiplication number sentence, a division number sentence or both?

Which representation do you find the easiest to understand? Why is this?

## What if?

Mrs Edwards asked Alice, Anushka, Matthew and Dylan to each show a different representation for this number sentence.

What four different representations might they have shown?

Try to think of some representations that are different from the ones above.

$$8 \times 2 = \boxed{\phantom{0}}$$

When you've finished, turn to page 80.

## Challenge

Each shape stands for a number. Shapes that are the same stand for the same number.
Work out the value of each shape, then complete the number sentences.

$\square + \square + \square = 9$   $12 \div \square = \bigcirc$

$\triangle \times \bigcirc = 8$   $\triangle + \triangle + \triangle = 6$

$\bigcirc \times \square = 12$   $\triangle \times \square = 6$

$8 \div \triangle = \bigcirc$   $\bigcirc + \bigcirc + \bigcirc = 12$

$\square \times \square = 9$   $6 \div \triangle = \square$

Explain how you worked out the value of each shape.

## Think about ...

Which number sentence will you start with to help you work out the value of the shapes? Why?

Once you have found the value of one shape, how can you use this to help you work out the value of the other shapes?

## What if?

Anushka says:

You can sort these number sentences into different pairs or groups.

How would you sort the ten number sentences into different pairs or groups?

Can you sort them in more than one way?

When you've finished, turn to page 80.

## Challenge

$$\frac{1}{4} \qquad \frac{1}{3} \qquad \frac{2}{4} \qquad \frac{1}{2} \qquad \frac{3}{4}$$

Sort these fractions into groups.

What's the same and what's different about them?

How many different ways can you sort the fractions, describing what's the same and what's different?

## Think about ...

Can you spot an **odd one out** – or even an **odd two out**?

Think about sorting the fractions into two, or even three, different groups.

## What if?

Finding $\frac{1}{4}$ is the same as dividing by 4.

How might Mrs Edwards show this to her class?

What might Mrs Edwards say about finding $\frac{1}{2}$ or $\frac{1}{3}$?

When you've finished, turn to page 80.

## Challenge

Matthew and Dylan have been asked to order these fractions, from smallest to largest.

$\frac{1}{3}$    $\frac{1}{4}$    $\frac{1}{2}$

$\frac{1}{2}$ is the smallest fraction because 2 is smaller than 3 or 4.

No, $\frac{1}{4}$ is the smallest.

Who's right? Explain why.

## Think about ...

You could use diagrams to explain who's right.

Think about finding $\frac{1}{3}$, $\frac{1}{4}$ and $\frac{1}{2}$ of a group of 12 objects. What do you notice?

## What if?

Alice says:

$\frac{3}{4}$ is larger than $\frac{1}{4}$.

Is Alice right? Explain why.

Anushka says:

$\frac{2}{4}$ is larger than $\frac{1}{2}$.

Is Anushka right? Explain why.

When you've finished, turn to page 80.

## Challenge

$\frac{1}{2}$ of 10 metres is 5 metres.

$\frac{3}{4}$ of 12 is 9.

$\frac{1}{4}$ of £8 is £4.

$\frac{1}{3}$ of 9 is 3.

Mrs Edwards says that one of the children is wrong.

Who's wrong?

Explain why.

## Think about ...

As well as explaining who's wrong and why, also explain why the other answers are correct.

Think about what you have to do to work out a fraction of a number.

## What if?

Mrs Edwards asks the children:

What would be your answer? Why?

Would you rather have $\frac{1}{2}$ of £12 or $\frac{3}{4}$ of £12?

When you've finished, turn to page 80.

## Challenge

Use the ruler in the picture to work out the length of each object.

Then use the < and > signs to compare the lengths of pairs of objects.

## Think about ...

Think carefully about those objects that don't line up with zero on the ruler.

Use the 'less than' and 'greater than' signs to compare objects as well as lengths.

## What if?

Alice says:

The pencil is 5 cm longer than the crayon.

Is Alice right?

Explain why.

Write some other statements that say how much longer or shorter one object is than another object.

When you've finished, turn to page 80.

## Challenge

The lemons on this scale all have the same mass.

What statements can you make about the mass of the grapefruit and the lemons?

If the grapefruit has a mass of 400 grams, what must be the mass of one lemon?

How do you know?

## Think about ...

Think about how the grapefruit compares to four lemons and also to just one lemon.

Can you compare the mass of a lemon and a grapefruit using the language of fractions?

## What if?

An apple is half the mass of a grapefruit.

If a lemon and an apple were placed on one side of the balance, and the grapefruit on the other side, what would the balance look like?

How do you know?

Can you draw a balance to prove it?

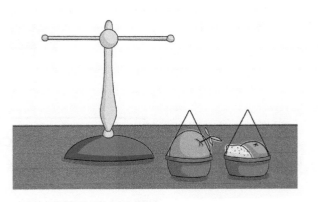

When you've finished, turn to page 80.

## Challenge

Who doesn't have half a litre of juice? How much juice do they have?

Explain why.

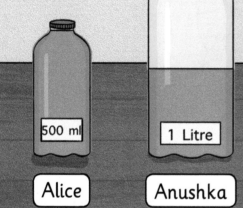

| | | | |
|---|---|---|---|
| 500 ml | 1 Litre | 2 Litre | 1000 ml |
| Alice | Anushka | Matthew | Dylan |

## Think about ...

What's different about all the bottles that have half a litre of juice in them?

The labels on the bottles show the amount of juice that each bottle holds when **full**.

## What if?

Matthew says: I have more juice in my bottle than anyone else because I have the largest bottle.

Alice says: No, I have more juice because my bottle is the only full bottle.

Who's right?

Explain why.

When you've finished, turn to page 80.

## Challenge

The following are Dylan's answers to a quiz about time.

What might the questions have been?

Time quiz

Dylan O'Herlihy

a) 60 ✓        e)

b) 12 ✓        f)

c) 24 ✓        g)

d) 30 ✓        h)

## Think about ...

Think about seconds, minutes, hours, days and months.

Could some answers have more than one question?

## What if?

These are the answers to the rest of the time quiz.

What might the questions have been?

Time quiz

Dylan O'Herlihy

a) 60 ✓        e) 7 ✓

b) 12 ✓        f) 31 ✓

c) 24 ✓        g) 365 ✓

d) 30 ✓        h) 15 ✓

When you've finished, turn to page 80.

## Challenge

There are only two different ways of making 67p.

You're both wrong. There are a lot more than two or five different ways of make 67p.

No. There are five different ways of making 67p.

Who's right?
Explain why.

## Think about ...

Think about using four coins, five coins, six coins, …

What is the best way to keep a record of all the different ways you can think of?

## What if?

Dylan says:

I can make 67p using only silver coins.

Is Dylan right?
Prove it.

When you've finished, turn to page 80.

## Challenge

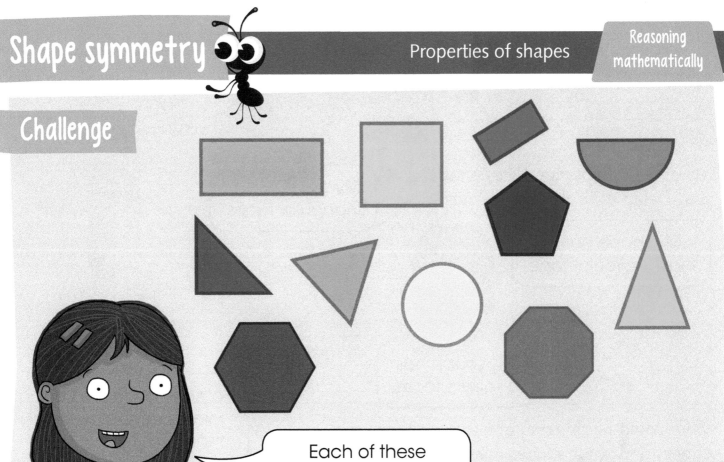

Each of these shapes has at least one line of symmetry.

Is Anushka right?
Prove it.

## Think about ...

Do any of these shapes have more than one line of symmetry?

Think carefully about how you're going to provide proof.

## What if?

Anushka also says: All of these shapes are 2-D shapes, so this means that they are also all polygons.

Matthew says: They are all 2-D shapes but they're not all polygons.

Who's right?
Explain why.

When you've finished, turn to page 80.

## Challenge

What 3-D shape are Alice, Anushka and Dylan each thinking about?

My shape has 4 triangular faces.

My shape has at least 1 circular face.

My shape has 6 square faces.

Some of you could be thinking about more than one 3-D shape.

Who could be thinking about more than one shape, and what shapes might these be?

Explain why.

## Think about ...

What is similar about some of the different 3-D shapes that you know?

Play '3-D shape odd one out' by thinking about which 3-D shapes are different, or 'the odd ones out'. Think about faces, edges and vertices, and flat and curved surfaces.

## What if?

Mrs Edwards played '3-D shape odd one out' with her class.

Matthew chose this shape as his 'odd one out'. Can you explain why?

Which shape would you choose in a game of '3-D shape odd one out' and why?

When you've finished, turn to page 80.

## Challenge

**HOME**

**START**

I programmed Ronnie to get home by walking forward 3 blocks, making a quarter turn anticlockwise and walking forward 3 blocks, then making another quarter turn anticlockwise and walking forward 2 blocks.

Is Dylan right? Explain why.

## Think about ...

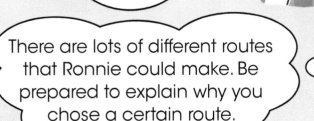

Think about using terms such as **quarter turn**, **one right angle**, **clockwise** and **anticlockwise**.

There are lots of different routes that Ronnie could make. Be prepared to explain why you chose a certain route.

## What if?

What is the shortest route you could programme Ronnie to walk home?

What if Ronnie has to stop off to refuel and then go to the mechanics on his way home?

When you've finished, turn to page 80.

## Challenge

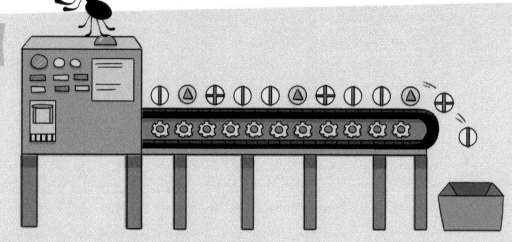

Draw the next screw to come out of the machine and travel along the conveyor belt.

What screw will come out of the machine after that?

Explain your reasoning.

## Think about ...

What repeating pattern do you notice?

Think carefully about whether you need to draw the screws coming out of the machine or falling into the tub.

## What if?

What if these screws are made in this pattern?

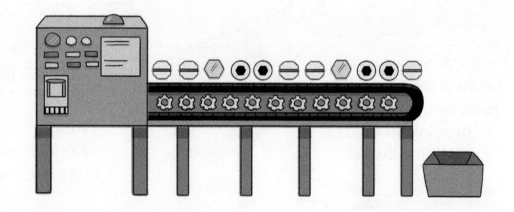

What screw has just fallen into the tub?

What screw fell into the tub before that one?

Explain your reasoning.

When you've finished, turn to page 80.

## Challenge

What is the same about these two graphs?
What's different?

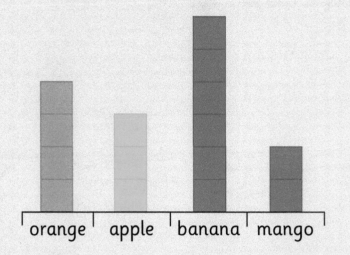

**Class 2A Favourite Fruit**

orange  apple  banana  mango

**Class 2B Favourite Fruit**

| mango | ☺ ☺ |
| banana | ☺ ☺ ☺ ☺ ☺ |
| apple | ☺ ☺ ☺ ☺ |
| orange | ☺ ☺ ☺ |

## Think about ...

Think about the data in the two graphs.

Also think about the way the data is displayed.

## What if?

What is the same about this table and the two graphs above? What's different?

| Class 2C Favourite Fruit | |
|---|---|
| mango | 2 |
| banana | 5 |
| apple | 4 |
| orange | 3 |

When you've finished, turn to page 80.

55

# Sorting dominoes

## Challenge

Which group of dominoes is the largest:

- the group with an **even** number of dots at both ends
- the group with an **odd** number of dots at both ends
- the group with an **even** number of dots at one end and an **odd** number of dots at the other end?

First make a prediction, then sort a set of dominoes into three groups to find out.

**You will need:**
- set of 6 × 6 dominoes

## Think about ...

Consider a blank as zero, and zero as an even number.

Even     Odd

Think about how you're going to record your results.

## What if?

What if you sort the dominoes into two groups:

- those with an even total number of dots
- those with an odd total number of dots?

First make a prediction, then find out.

What do you notice about your results?

How else could you sort the dominoes?

When you've finished, turn to page 80.

# Visitors

## Challenge

Predict, and then record, how many visitors come into your classroom in a day.

Who are the visitors? Are they children, teachers, other school staff members, parents or other people?

Also, find out their reason for visiting and whom they have come to see.

## Think about ...

Think about how you're going to collect, record and organise the information.

Think about how to display your information in the clearest way possible.

## What if?

Keep a record of all the different places you visit in one week.

You visit your school five times in one week. What other places do you visit?

When you've finished, turn to page 80.

## Challenge

My phone number is 01632 960872. This is a total of 44.

Add together all the digits that make up your phone number.

Who has the largest total in your group?

Who has the smallest total?

Order the totals, starting with the largest.

## Think about ...

Are you going to use home phone numbers or mobile numbers?

Will you include the area code or mobile phone code?

For What if?, you'll first need to complete the Name Phone Number Code for letters I to Z.

## What if?

What is your name phone number?

Work out some name phone numbers of your friends.

My name phone number is 112935.

ALICE
11293 5

**NAME PHONE NUMBER CODE**

| | |
|---|---|
| A = 1 | B = 2 |
| C = 3 | D = 4 |
| E = 5 | F = 6 |
| G = 7 | H = 8 |

When you've finished, turn to page 80.

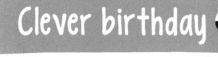 # Clever birthday

## Challenge

I can make all the numbers from 1 to 20 using just the digits in my date of birth and the signs +, −, ×, ÷ and =.

Alice was born on 15th April 2011, so the digits that make up Alice's date of birth are: 1, 5, 4, 2, 0, 1, 1.

Is Alice right?

Prove it.

## Think about ...

Hint: One way of making the number 12 is:
$2 \times 5 + 1 + 1$.

You can use any of the seven digits in Alice's date of birth. But you can only use them once in each calculation. So, although there are three 1s, there is only one 2.

$12 = 2 \times 5 + 1 + 1$ ✓
$12 = 2 \times 5 + 2$ ✗

## What if?

Can you make all the numbers from 1 to 20 using your date of birth?

When you've finished, turn to page 80.

# Phone pad patterns

## Challenge

Investigate the patterns of digits on a phone.

Add the three digits in each column.

Add the three digits in each row.

Add the three digits in each diagonal.

What do you notice?

## Think about ...

Look at the sum of each column and row.

Can you spot more than one pattern for the totals of the columns, rows and diagonals?

## What if?

Take a 2 by 2 square of digits on the phone and add each pair of diagonal digits. What do you notice? Now take a different 2 by 2 square and do the same thing. What do you notice?

What if you find the difference between pairs of diagonal digits?

What if you multiply them?

When you've finished, turn to page 80.

# Calculating dominoes

## Challenge

**You will need:**
• set of 6 × 6 dominoes

Choose a domino.

Add together the number of dots at each end of the domino.

Record your results.

Now do this for all the other dominoes.

What is the most common total?

## Think about ...

Before you start, can you predict which total (+), difference (−) and product (×) will be the most common?

Think about how you're going to record your results.

## What if?

What if you find the difference between the number of dots at each end of a domino?

What is the most common difference?

What if you only choose dominoes with dots at both ends (in other words, no blanks) and multiply the number of dots at one end of the domino by the number of dots at the other end?

When you've finished, turn to page 80.

What are the most common products?

# Stamps

## Challenge

### 1st and 2nd Class stamps

### 'Make up value' stamps less than £2

The most common types of stamp in the UK are 1st and 2nd Class stamps.

What is the value of the four different 1st and 2nd Class stamps?

Mrs Edwards has six letters to post. The costs of the six letters are:

£1.22     £1.74     £1.30     £1.58     £1.66     £1.86

Work out which stamps Mrs Edwards needs to buy to post each of her letters.

## Think about …

Use the smallest number of stamps for each letter.

Use different combinations of 1st and 2nd Class stamps and 'Make up value' stamps.

## What if?

Work out a different way that Mrs Edwards could pay for each letter. Try to use as few stamps as possible.

When you've finished, turn to page 80.

# Body links

## Challenge

Join the tip of your middle finger to the tip of your thumb. Is this the same size as around your wrist?

**You will need:**
- measuring equipment

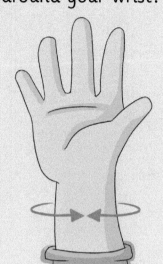

What do you notice about:

- the size of your wrist and the size of your thumb?
- the size of your index finger and the distance from the tip of your nose to your chin?

## Think about ...

What type of measuring equipment will you need to make accurate measurements?

Think carefully about how you're going to measure around your wrist, thumb, neck, waist and fist.

## What if?

What do you notice about:

- the size of your neck and the size of your wrist?
- the size of your neck and the size of your waist?
- the size of your fist and the length of your foot?

Find other parts of your body that are the same length, or where one part is twice the length of the other.

When you've finished, turn to page 80.

# Walking

## Challenge

**You will need:**
- measuring equipment

How many steps would it take you to walk around the school grounds?

How long would it take you?

Estimate first, then find out.

## Think about ...

It may not be possible to get an exact answer, so think about how you can come up with a **good approximation**.

Think about how you're going to keep a record of how many steps you take and how far you walk.

## What if?

How far do you walk in a day?

How many steps is this?

When you've finished, turn to page 80.

# Holding hands

## Challenge

**You will need:**
- measuring equipment

Approximately how many children, holding hands, would you need to reach all the way around the school?

Estimate first, then find out.

Write about what you did.

## Think about ...

It probably won't be possible to get enough children to hold hands all the way around the school, so think about how you can come up with a **good approximation**.

Think carefully about what you will need to use to find out.

## What if?

How far is it from one side of the playground to the other?

Estimate first, then find out.

Write about what you did.

When you've finished, turn to page 80.

# Balls

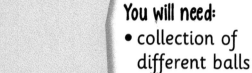

## Challenge

**You will need:**
- collection of different balls
- measuring equipment

Without lifting the balls, order them from the one you think is the lightest to the one you think is the heaviest.

Now see if you were right.

How heavy is each ball?

## Think about ...

Think carefully about how you're going to record your results.

How are you going to measure the height of each ball in the moment of its greatest bounce height?

## What if?

How high does each ball bounce?

Estimate first, then find out.

Order the balls from the one that bounces the lowest to the one that bounces the highest.

When you've finished, turn to page 80.

# Cooked and uncooked

## Challenge

Does a boiled egg have a greater mass than a raw egg?

Estimate first, then find out.

Write about how you found out.

**You will need:**
- a boiled egg and a raw egg
- 50 g uncooked rice and 50 g uncooked rice that is then cooked
- 50 g uncooked pasta and 50 g uncooked pasta that is then cooked
- measuring equipment

## Think about ...

What measuring equipment will you use?

What units of measure will you use?

## What if?

Does cooked rice have a greater mass than uncooked rice?

If so, by how much?

What about cooked and uncooked pasta?

When you've finished, turn to page 80.

67

# Tinned fruit

## Challenge

In a tin of fruit, how much is fruit and how much is juice?
Write about how you found out.

**You will need:**
- tins of fruit
- measuring equipment

**Peaches**

**Pineapple**

**Cherries**

## Think about ...

What measuring equipment will you use?

What units of measure will you use?

## What if?

Is it possible to say that there is more juice than fruit, or more fruit than juice in a tin?
Explain why.

When you've finished, turn to page 80.

# Orange drink

## Challenge

A plastic cup holds 200 ml.

A bottle of orange squash is 1 litre.

To make an orange drink you mix 1 part squash to 4 parts water.

**You will need:**
- 1 litre bottle of orange squash
- 200 ml plastic cups
- water
- measuring equipment

How much **orange squash** would you need to use to give everyone in your group a cup of orange drink?

How much **water** would you need to use to give everyone in your group a cup of orange drink?

## Think about ...

How many millilitres of orange squash do you need for each cup?

How many millilitres of water do you need for each cup?

## What if?

Mrs Edwards says:

You should drink at least 1 litre of liquid a day.

How much liquid do you drink in a day?

Estimate first, then find out.

Write about what you did to find out.

When you've finished, turn to page 80.

# TV times

## Challenge

Look at the starting times of television programmes.

Investigate how many programmes start at:

- o'clock times
- half past times
- quarter past times
- quarter to times.

Which is the most common time for a programme to start?

Why do you think this is?

**You will need:**
- television guide

## Think about ...

Is it the same at weekends as it is on weekdays?

Are starting times different for different channels?

## What if?

How long do most programmes run for – one hour, half an hour or other lengths of time?

What can you say about all the programmes that run for one hour? Is there something similar about them?

What about all the programmes that run for half an hour or other lengths of time?

When you've finished, turn to page 80.

# Paper clips

## Challenge

How many paper clips can you link together in 1 minute?

Estimate first, then find out.

Once you've worked out how many paper clips you can link together in 1 minute, estimate how many you could link together in:

- 2 minutes
- 5 minutes
- 10 minutes.

Now find out how close your estimates were.

**You will need:**
- lots of paper clips
- minute timer or wall clock with a seconds hand
- measuring equipment

## Think about ...

Think about the best technique to use to link as many paper clips together as possible. You might want to have a 'trial run'!

If your estimates are not the same as the actual number of paper clips you linked together, can you think why this might be?

## What if?

Make a paper clip bracelet and necklace.

What is the cost of your bracelet and necklace, if the cost of one paper clip is 2p?

When you've finished, turn to page 80.

71

## Challenge

Mrs Edwards has £12.

She wants to find all the different ways she can share the £12 between 2, 3 and 4 children.

She records each way in a table.

How many different ways can you find of sharing £12 between 2, 3 and 4 children?

Record your results in a table similar to Mrs Edwards' table.

Are all Mrs Edwards' ways of sharing £12 fair?

Explain why?

| Number of children | Fraction | Money |
|---|---|---|
| 2 | $\frac{1}{2}+\frac{1}{2}$ | £6 + £6 |
| 3 | $\frac{1}{2}+\frac{1}{4}+\frac{1}{4}$ | £6 + £3 + £3 |
| | | |
| | | |
| | | |

## Think about ...

Think about the different fractions you can use to share the £12. Which fractions are 'good fractions' to use?

Think about how this model may help.

## What if?

What if Mrs Edwards has £24 to share between 2, 3 and 4 children?

Compare your results for sharing £24 with your results for sharing £12. What do you notice?

When you've finished, turn to page 80.

# £100 to spend!

## Challenge

**You will need:**
- toy catalogue

Look through a toy catalogue. It may be a printed catalogue or an online catalogue.

You have £100 to spend.

What toys would you like to buy? How much do they cost?

## Think about ...

Can you spend exactly £100?

If you can't spend exactly £100, can you work out how much of your £100 you have left?

## What if?

Petunia £20

Fuchsia £15

Geranium £25

Busy Lizzie £12

Begonia £18

Mrs Edwards has £100 to spend on flower baskets for around the school. She can choose from these five baskets.

What flower baskets should she buy?

When you've finished, turn to page 80.

# Supermarkets

## Challenge

Investigate the different shapes of packaging that are used for food sold in supermarkets.

Which shapes have symmetry?

Which two shapes are the most popular for packaging? Why do you think this is?

**You will need:**
- food magazines or shopping flyers
- scissors

## Think about ...

Is there a difference in the way dry food and wet food are packaged?

Think carefully about how you're going to show a line of symmetry in different fruit and vegetables. You could find photos in magazines or shopping flyers.

## What if?

Which fruit and vegetables have symmetry?

Can you show where their lines of symmetry are?

Which fruit and vegetables have no symmetry?

When you've finished, turn to page 80.

# Animal shapes

## Challenge

What is your favourite animal?

Using only these shapes, draw your favourite animal.

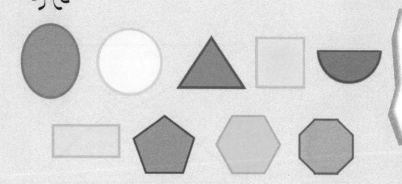

**You will need:**
- set of 2-D shapes
- a friend

You can use as many of each shape as you like and you don't have to use them all.

Keep your drawing secret from your friend.

Now describe your animal to your friend and ask them to draw it.

Does your friend's drawing look like yours?

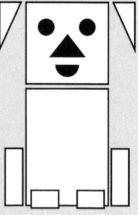

## Think about ...

You can use different sizes, orientations and types of the same shape. For example, you could draw a triangle in any of these ways (and more):

It might be a good idea to make some sketches first.

## What if?

The picture of the dog above was drawn using 12 shapes.

What other pictures can you draw using just 12 shapes? Your picture doesn't have to be of an animal.

When you've finished, turn to page 80.

# Envelopes

## Challenge

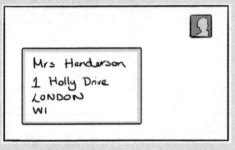

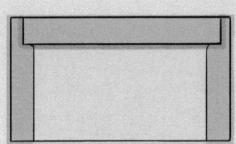

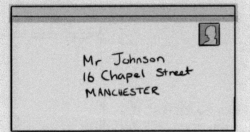

**You will need:**
- several sheets of A4 paper
- square sheet of paper
- scissors
- glue
- sticky tape
- ruler

Design and make the largest envelope you can from a sheet of A4 paper.

When you're happy with your design, draw a template of your envelope so that someone else can make exactly the same envelope.

## Think about ...

Use **trial and improvement** until you think you've designed the largest envelope possible.

Your fold down flap can be either triangular or rectangular.

## What if?

Can you make an envelope just by folding a sheet of paper and without making any cuts?

You can use either a square sheet of paper or a rectangular sheet of paper, such as A4 size.

When you've finished, turn to page 80.

# Apple patterns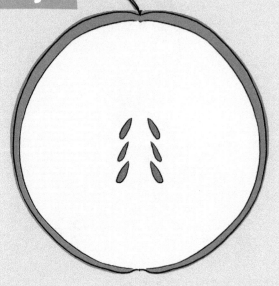

## Challenge

**You will need:**
- an apple cut in half and one of the halves cut into quarters
- paint in two different colours
- two paintbrushes
- large sheet of paper

Using the flat face of half an apple and one of the flat faces of a quarter of an apple, investigate the different repeating patterns you can make using one paint colour.

## Think about ...

Make sure that you include enough apple prints in your pattern so someone can spot your pattern and continue it.

Make sure that your pattern repeats at least once.

## What if?

Investigate the different repeating patterns you can make using two different coloured paints.

Show all of your patterns to a friend and ask them to continue your patterns.

When you've finished, turn to page 80.

77

# Breakfast

Using and applying mathematics in real-world contexts

## Challenge

What did everyone in your class have for breakfast this morning?

What did each person have to eat?

What did each person have to drink?

What is the most popular breakfast in your class?

Display your results.

## Think about ...

How are you going to collect and organise your results?

What is the best way to display your results – perhaps using a table, diagram or chart?

## What if?

If tomorrow you had to give breakfast to everyone in your class, what would you need to buy?

Approximately how much of each thing would you need to buy?

When you've finished, turn to page 80.

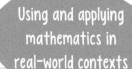
## Challenge

"We're going to get a class pet. What type of pet shall we have?"

Find out what type of pet most children in your class would choose, and why.

Present your results in a table, diagram or chart so that you can convince Mrs Edwards of the pet your class should have.

## Think about ...

Think about what pet is most suitable for a classroom. It might be different from a pet you would have at home.

Think about how you are going to collect, organise and present your results.

## What if?

Find out from the children in your class what name you should give your class pet.

Again, present your results in a way that will convince Mrs Edwards of the name your pet should have.

When you've finished, turn to page 80.

**Share** Share your results.

**Discuss** Discuss any results that are different.

Which result is correct?

Might there be more than one solution?

**Share** Share the methods used.

**Discuss** Discuss the similarities and differences in the methods used.

Which method worked best?

Are there any other ways to go about solving the problem?

**Share** Share what you have learned.

**Discuss** Discuss what you would do the same and what you would do differently next time.

Is there anything you would do differently?

What have you learned for next time?

What would you do the same?